Composite Structures & Constructio
Wet Lay-up & Prepreg Construct
Automotive / Marine Applications

Spencer G. Gould

Table of Contents

Prologue: **6**

Chapter 1 - Shop Safety **10**

   1.1 - Safety Equipment **10**

Chapter 2 - Tools **18**

   2.1 - "The Easy Squeegee" **19**

   2.2 - Small Short-Nap Rollers **20**

   2.3 - The 1" Brush..**21**

   2.4 - Rotary Razor "Pizza" Cutter **23**

   2.5 - Scissors **24**

   2.6 - Epoxy Pumps **26**

   2.7 - Pre-Marked Cups for Volume Mixing **32**

   2.8 - Shop Vacuum **34**

   2.9 - Laminating Table **35**

   2.10 - Partial Molding Table **36**

   2.11 - Band Saw...**38**

   2.12 - Rotary Tools **39**

Chapter 3 - Raw Materials **41**

   3.1 - Reinforcements: Fiberglass, Carbon, and Kevlar..........**41**

      3.1.1 - Fiberglass **42**

      3.1.2 - Carbon Fiber **49**

      3.1.3 - Kevlar **51**

      3.1.4 - Pultruded Materials **52**

- 3.1.5 - Prepreg Materials ... 55
- 3.1.6 - Peel Ply ... 58
- 3.1.7 - Release Film ... 58
- 3.1.8 - Vacuum Bagging Film ... 58
- 3.1.9 - Transfer Plastic ... 59

3.2 – Resin Systems ... **61**
- 3.2.1 - Epoxy Resin Systems ... 61
- 3.2.2 - Vinyl Ester Resins ... 64
- 3.2.3 - Polyester Resins ... 64

3.3 - Cores ... **65**
- 3.3.1 - Foam Cores ... 65
- 3.3.2 - Honeycomb Core ... 69
- 3.3.3 - Wood Cores ... 72

3.4 - Inserts & Hard Points ... **73**
- 3.4.1 - G10-FR4 ... 74
- 3.4.2 - Phenolic ... 77
- 3.4.3 - Carbon Sheet ... 79
- 3.4.4 - Wood ... 81
- 3.4.5 - Aluminum ... 81
- 3.4.6 - Steel ... 82
- 3.4.7 - General Metal Prep for Inserts, Bushings & Jacketing ... 82

3.5 - Fillers ... **83**
- 3.5.1 - Micro Balloons ... 84
- 3.5.2 - Flox ... 85

- 3.5.3 - Milled Glass Fibers 85
- 3.5.4 - Cab-o-sil 86
- 3.6 - Adhesives 87
  - 3.6.1 - Epoxy Adhesives 87
  - 3.6.2 - Cyanoacrylate 90
  - 3.6.3 - Polyurethanes 91

Chapter 4 - Gigs, Fixtures, Molds, & Moldless Methods 92
- 4.1 - The Value of Tape and Single-Use Foam Molds 92
- 4.2 - MDF 95
- 4.3 - Partial Molding Techniques 96
- 4.4 - Vacuum Bagging 100
- 4.5 - Hot Wire Foam Cores 102

Chapter 5 - Cloth Kits 104
- 5.1 - Layout 104
- 5.2 - Cloth Cutting: Methods Precautions and Considerations 105
- 5.3 - Kitting 106
- 5.4 - Back Rolling 107

Chapter 6 - The Game Plan 108
- 6.1 - The Checklist 108
- 6.2 - The Working Environment 109
- 6.3 - Preempting, Eliminating, and Destroying Distractions! 109
- 6.4 - Composite Drawings & Layup Instructions. 113

Chapter 7 - Bonding 116

    7.1 - General Alignment 116

    7.2 - Cleco Alignment 120

    7.3 - Bonding Prep 122

        7.3.1 - Peel Ply Removal 122

        7.3.2 - Mechanical Roughing Up. 124

    7.4 - Bonding Trowel 130

Chapter 8 - Finishing 131

    8.1 - General Game Plan 131

    8.2 - Filling-in the Weave Options 131

    8.3 - UV protection 132

Chapter 9 - Additional Resources 133

About the Author 138

Prologue:

This is book is intended as a guide for people at all skill levels. From beginners all the way through highly proficient composite layup technicians, shop foremen, designers, and engineers, this book has something for everyone.

I went through the white-collar education and career path of Aerospace / Mechanical Engineering, but have found great value building my competency by learning blue-collar skills like composite layups, welding, machining, and woodworking. Time and time again, I have found that the most valuable co-workers have hands-on skills with auto, aero, and/or civil endeavors outside of their day jobs.

Within the scope of this book, the word "Composites" primarily refers to Polymer Matrix Composites (PMC). These materials consist of fiberglass/carbon face sheets, foam/honeycomb cores, and epoxy based "matrix" resin and adhesive systems.

So why composite materials? The answer is simple: general ease of use with nearly unlimited contouring options. You can achieve incredibly light-weight/high-strength structures, with the wildest contours you can dream up in your garage, with a minimum investment in tools. This was demonstrated by Burt Rutan, Glasair, and Lancair in their early days.

At its most basic level, working with composites is an additive process; meaning you only add the material you need to get the job done. This is one of the key principles of how composite materials achieve excellent strength to weight ratios. In comparison, CNC metal hog-outs and sheet metal construction are "subtractive" in nature. Additionally, with theses subtractive manufacturing processes, you must use large and expensive equipment. You will expend considerable time and effort to remove the deadweight in a metal structure. While weight and cost savings are critical items in anything that flies, it also applies to cars and boats as well.

In addition to this book (in either paperback or Kindle formats), I have a companion web page that covers composite materials and fabrication equipment in greater detail, as well as including links to projects. www.gouldaero.com/pmc can provide you with an additional resource for years into the future.

This book is dedicated to my parents who let me find and follow my passions in life, and helped support my aviation addiction.

Disclaimer: Always read, understand, and follow all applicable safety procedures and methods for everything covered in this book; including safety information that comes with your supplies and equipment. The author/publisher assumes <u>NO</u> liability.

Chapter 1 - Shop Safety

1.1 - Safety Equipment

During the course of your project, there will be many opportunities for injury and accumulative buildup of chemicals and dust in and on your body. You should take every action to protect yourself and your helpers. A drop of epoxy landing on your safety glasses may be a minor inconvenience, but imagine what would happen if you were not wearing them! Remember to be safety conscious.

The chemicals that comprise laminating resins can be harmful to the body if proper precautions are not taken. An allergic reaction called "sensitization" occurs when you become highly allergic to a particular resin. Keeping the resin off your skin and vapors out of your lungs can keep this from happening.

Impact-resistant safety glasses should be worn any time chemicals are poured, mixed, laminated, or bonded. They should also be worn during any cutting, sanding, or grinding operation on cured laminates. The glasses should have good coverage in front of your eyes and have protection from underneath and the sides.

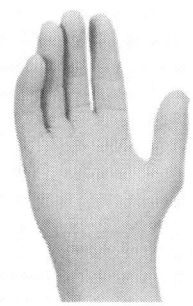

The majority of resin contact with your skin will happen on your hands. I recommend disposable gloves. I am not a fan of, or believer in, "skin barrier creams". My number one choice is disposable nitrile gloves. A box of 100 can be purchased at Harbor Freight for less than 10 bucks, and these gloves hold up well. Nitrile gloves are highly resistant to chemicals and inexpensive. They are also the go-to for people who are allergic to vinyl gloves. Another reason to choose nitrile over vinyl gloves is that vinyl gloves can become soft and absorb resin during a layup. Using disposable gloves is cost effective, as cleaning the resin off the gloves is not worth your time or money.

Proper ventilation is necessary to remove out-gassing fumes from your working area. Some laminating resins have a very powerful and detectable smell, while some may have no significant smell at all during layup and the curing process. At minimum, make sure you can put a reasonably sized shop fan at an open door or window to exhaust the fumes.

Heating your resin in a heat box, or with a heat lamp, is a way to make the resin thinner (lower the viscosity) and/or bring it up to working temperature in cold environments. However, warm or hot resin can outgas chemicals! Typically, this can be seen as a white mist that is lighter than air and accumulates at the ceiling in an unvented room. Trust me, you don't want this stuff in your lungs. If you see a mist-cloud form near the ceiling while laminating, STOP and crank up the ventilation. This can be a challenge in colder environments, as most laminating resins work best at a temperature of 70°F or higher. In this case, a heater/fan can be used to bring in air from the outside, and a normal fan can be used to exhaust the workspace. I have found that using a roller to laminate will yield better results in general, but is especially helpful for working with cold high-viscosity resin.

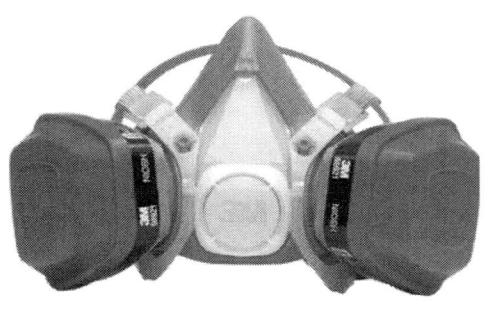

Coupled with good ventilation, a NIOSH-rated respirator equipped with organic vapor cartridges should be worn for any laminating and bonding operations involving epoxy, vinyl ester, and polyester products. These are all VERY POWERFUL chemicals in their liquid form, and outgas while curing. Human lungs are very difficult to replace! I speak from first-hand experience losing a family member to complications associated with a double lung transplant after a lifetime of chain smoking. Most brick-and-mortar home improvement stores and online-retailers like Amazon carry heavy-duty respirators. I would recommend the 3M 65021HA1 model, or the latest version of it. It costs under 30 bucks, and is cheaper than a co-pay to see the doctor.

Protective clothing is a necessity. Make sure to have a set of dedicated shop clothes. Dust from composites will work its way into your clothes as you work. Make sure to dust them off and wash them regularly. Dried epoxy globs on nice cloths may not bother you, but it can severely anger mothers, girlfriends, and wives when they show up on your Sunday's finest. Draw your own conclusions on how that can affect shop safety (arguments, objects thrown at you, temporary residence in a doghouse, etc.).

1.2 - Additional Safety Considerations

Avoid trip hazards and "head-hitters" in your work environment. Before you laminate, make sure you can access the entire part you are laminating without any objects in your way that may cause you to trip or hit your head. Cups of resin can go flying in the air and productivity can be severely reduced if you fail to anticipate potential hazards.

Always protect your cloth stock and cloth kits with plastic sheets, bags, or boxes when not in use. Keeping the cloth in a low-humidity environment for long-term storage is also strongly advised. Cloth needs to be clean and free of oil for safety and maximum laminate performance. Do not attempt to use cloth with visible oil stains, or that has been packed with dust. The consequences are not worth it. When cutting out cloth for a laminate, make sure to use gloves. The natural oils on your hands will transfer to the cloth and reduce its final performance characteristics.

Resins should be sealed and stored indoors when they are not in use. Out-gassing can occur with open containers. If you are using a resin transfer pump, make sure its reservoirs are covered.

Core material is easily damaged and should be stored covered and indoors, away from high-traffic areas. Many core materials are UV-sensitive. Exposure to sunlight can cause them to yellow and degrade. Keeping core materials away from windows and vents where sunlight can penetrate will preserve your material supply.

Chapter 2 - Tools

Wet lay-up requires a dedicated set of tools. However, it is important to note that some tools are disposable. Cleaning tools that can be thrown away takes time away from building. Brushes and short-nap rollers are tools that fall into in this disposable category.

www.gouldaero.com/pmc has additional resources where you can buy some of these tools.

2.1 - "The Easy Squeegee"

Vendors that sell composite materials love to sell squeegees costing several dollars apiece. That cost makes you feel compelled to clean them. What if you could have hundreds of disposable squeegees for the cost of a couple of those fancy squeegees that work just as well? The best part is they require no cleaning. It's simple, your local home improvement stores sell extra vertical blinds (typically in a shrink-wrapped package). A single package should last for a complete build. Just cut the individual blind slats to length with scissors or a chop saw before use.

2.2 - Small Short-Nap Rollers

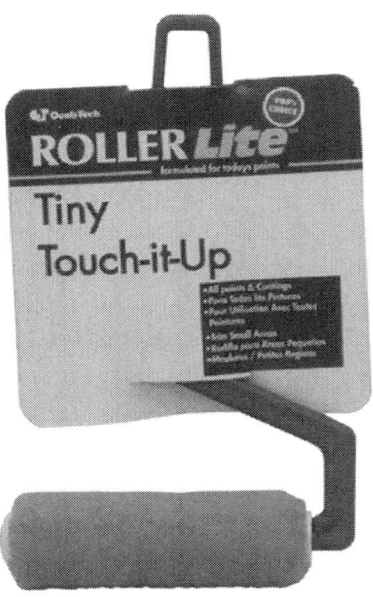

By far the best large-area lamination technique I have learned is using rollers to do the bulk of the laminating. The smaller the size of the roller's diameter, the more pressure you can put on the laminate. Short-nap rollers absorb less epoxy than those made of foam. They are not free, but the benefits far outweigh the cost. They will produce a laminate with higher quality, optimum resin-to-fabric content, 100% contact with the core, and it is faster than using a brush. If the cloth lifts up onto the roller, the cloth is not making contact with the core. This lets you know where you need to go back and make sure the cloth sticks to the core with more roller passes and/or more resin. This quality and safety feature is not available with other lamination methods such as using brushes, squeegees, or your hands! You can do multiple small lay-ups in one job to help offset the cost of the roller.

2.3 - The 1" Brush.

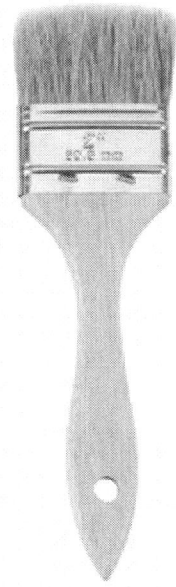

Many disposable brushes will be required through the course of a project. These are sometimes called chip-brushes. The typical home improvement store will have a significant markup on disposable brushes. Remember the old adage, "Watch the pennies, and the dollars watch themselves." Save money by purchasing brushes in bulk through mail order or at stores like Harbor Freight Tools. Keep the brushes covered and clean while in storage. After use, I keep the brush around until the epoxy has fully cured, and then I throw it away for an extra quality control check. If the epoxy has not cured on the brush, I know there is a problem that could affect the structural integrity of the part I was laying up. Harbor Freight is a great source for these brushes, and the most frequently used sizes and part numbers are listed here. 1" x 36 pack = #61491 and 2" X 36 pack = #61493. Remember your up-to-date 20%-off coupon before you go to the store, and watch for promotional sales.

2.4 - Rotary Razor "Pizza" Cutter

Rotary razors are the go-to tool for cutting Fiberglass and Carbon fiber cloth. Typically made by Fiskars, they can be purchased through Michaels Crafts or online. Item number 10221519.

2.5 - Scissors

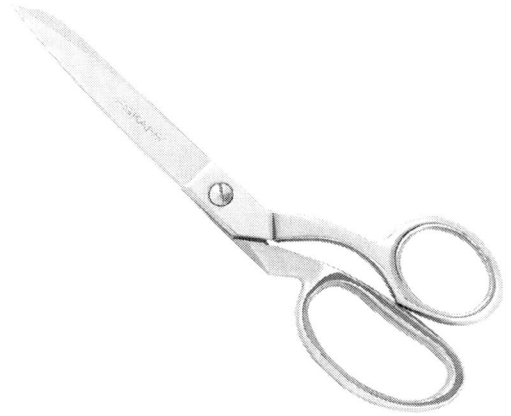

You will need high quality scissors to work with cutting dry and wet fiberglass and carbon fiber. It is best to designate one set for dry cutting and one for wet cutting. I recommend models that are solid metal and chrome plated as these are "keeper" tools, and you may need to use heavy-duty solvents to clean them up. These solvents can melt or embrittle plastic components on the scissors. Michaels Crafts is a good brick and mortar store source. They have coupons available in fliers, newspapers, and online. Fiskars Premier Forged Scissors is my go-to model for these purposes. Fiskars products are also available through Amazon.

I recommend keeping a roll of paper towels handy when you are laminating for cleaning up spills and to blotting up excess epoxy that cannot be taken care of with a squeegee.

2.6 - Epoxy Pumps

Epoxy pumps are a productivity improver for anyone doing high volume composites work. Scales can be tedious, time-consuming, and are more likely to produce an error. However, some epoxy systems only list volume mixes, and some people still prefer weight mixing. Remember: every minute spent on an inefficient measuring process is one less minute you have for productive building.

It is wise to find a variable ratio pump as opposed to a fixed one. A variable ratio pump will allow you to use a variety of lamination systems. Another wise investment is a scavenge system to catch the resin and hardener in separate cups so it can be returned to the reservoir, preventing wasted epoxy. A pump with a beveled edge that points away from each tube can give an exact drip location. This helps keep part A and part B separate.

Epoxy pumps can be very expensive when purchased new, but they can be worth the money if you are doing high-volume work. Another option is to purchase a used pump. Some builders may give away, sell cheap or long-term-lend their epoxy pumps after they are done with their composite project. This is another benefit of belonging to a local EAA chapter.

Used epoxy pumps are usually extremely messy and may have been sitting unused for years. They may have a different brand of epoxy in the reservoirs, or old goop inside. After a certain amount of use, epoxy pumps need to be rebuilt. Springs, valves, ball bearings, and lines all gum up. To play it safe, a full rebuild should be done. After a thorough disassembly, you will observe that some parts can be cleaned, while others are easier to throw away. So, it's time to break out the heavy-duty gloves, respirator, and cleaning agents. Acetone and "Aircraft Remover" work well for cleaning up the metal parts. Keep this in mind when someone recommends using acetone to prep fiberglass parts (this is akin to prepping metal with sulfuric acid). If the reservoirs are made of metal, they are worth the effort. For plastic reservoirs, cleaning agents powerful enough to clean epoxy will often dissolve the plastic. I have found the polypropylene bottles (look for the number 5 recycling symbol) often work well as reservoirs and last for years of service. High-density polyethylene (number 2) can also work, but will show some degradation over time. The best option for sourcing tanks on a re-build is to check Wal-Mart, Target, Container Store, IKEA, etc. for stainless steel containers that are the right size and shape.

The lines that run from the cylinder manifold to the business-end of the pump also need attention during a rebuild. I would recommend aluminum or stainless steel for the replacement lines. I have seen some scary stuff come up when it comes to the lines. Long story short, the epoxy hardener is highly reactive with many commonly available tubing materials. Clear vinyl tubing scales up, and ends up as a contaminant in your fresh batch of epoxy. The degradation takes a few months, but it becomes obvious when it happens. Another material to avoid is copper tubing, despite how tempting it may seem because of its easy-to-form nature. I discovered that epoxy hardener reacts with the copper resulting in a drastic color change in the catch-cup only a few hours after the first use. The normally amber hardener turned into blue-grey drips on the other side. You may not observe these effects with normal pumping, but if you scavenge the drips, the scavenged material will contaminate your hardener. Note: A color change is proof of a chemical reaction altering the chemistry of the epoxy.

Initial and regular calibration of your epoxy pump is critical for strength and safety. Graduated cups can be used for this. Make sure you strictly adhere to the ratio prescribed by the epoxy vendor. When using an epoxy pump, always do a "reality check". If you are supposed to have a 3 to 1 resin to hardener ratio, and you see a 50:50 ratio in the cup, it's time to check out what is going on.

Here is a picture of a 3:1 epoxy calibration from the pump. Tick lines were already added on the cups before pumping the epoxy (emphasized with arrows below).

It's wise to keep your epoxy cups for a week or so. Scratch the surface with a nail, if the residual hardened epoxy comes off white and chalky, you're good. If it's not cured (still gooey), it's time to do some investigating so you can adjust your mix ratio.

Some epoxy kits come with their own light-use built-in pumps that look similar to a ketchup pump at a fast food restaurant. West System is one of the main epoxy manufactures that uses this system, and it can be low cost and effective.

2.7 - Pre-Marked Cups for Volume Mixing

Another alternative I've found that works well is pre-marked cups. You can purchase them at low cost and in bulk from Amazon. If you're doing layups in the 1-2 square foot range, 30 ml medicine cups work out well.

I usually use PTM&W's Aero-poxy that's mixed in a ratio of 3 parts resin to 1 part hardener. A typical mix has a total volume of 20 ml, so I make a sharpie mark at 5 ml and at 20 ml. I pour the hardener to the 5 ml line. The hardener's viscosity is much lower than the resin, and results in a lower meniscus all around. This provides higher mix accuracy. I then pour the resin on top of the 5 ml of hardener until the total volume is at the 20 ml line. To explain the math a bit more: The difference between 20 ml and 5 ml is 15 ml, 15:5 = 3:1. For larger batches, I use 9-ounce clear plastic cups from Sam's Club. I've made a template that has graduated marks for a variety of volumes, all with the same mix ratio. This template goes inside the cup and I sharpie mark the applicable lines onto the outside of the cup. Note: These cups are tapered, so you can't just use a ruler to measure volume because cup height is not linear with respect to volume in a tapered cup. Of course, you can do some fancy volume of a cone math, but I recommend taking the simple route of using the medicine cups and pouring in increments of water into the 9-ounce cup and sharpie marking all the various levels. Four 20 ml pours (each with its own tick mark) could give you a 20 ml and 80 ml combo. You can apply this for all kinds of other numbers 5 and 20, 10 and 40, 40 and 160...

2.8 - Shop Vacuum

A shop vacuum with a dedicated brush attachment is a necessity. When prepping a core, or a bond site, a dust-free surface is critical. Make sure you know the history of the brush attachment. If it was used to clean up soil or oil, it could contaminate the aircraft-grade composite part or core material you are trying to work on.

2.9 - Laminating Table

A melamine-covered MDF or particleboard sheet makes for an excellent laminating table base. I cover the laminating table with a new party tablecloth before each day's laminating. If any epoxy drops get onto the base table, they can be scraped off the melamine surface with a razor blade.

2.10 - Partial Molding Table

A partial molding table is intended to be used for non-vacuum bagged composite parts, where the laminate must make a transition from a foam or wood core to a face sheet or face sheet laminate. The key to making a partial molding table is to use something that's easy to drive 4-penny flush head nails into. These will hold the core down against the table (keeping the core in a constant position and holding the beveled edges tight to the transition zone).

4-Penny flush or reduced head nails are the ideal way to pin a foam core against a dedicated table. The small size of the head pulls right through the core when it is time to de-mold the part. Once you are ready to laminate the other side of the part, the holes can be filled in with micro-balloons before lamination.

2.11 - Band Saw

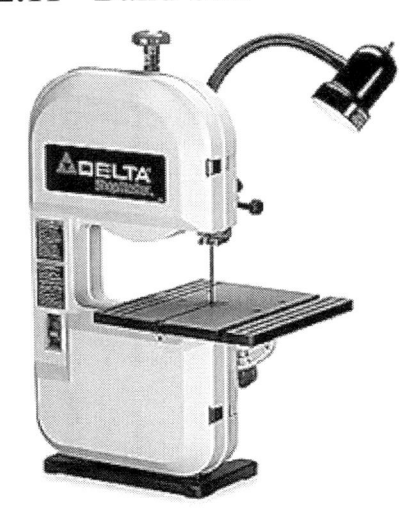

A band saw will be one of your main tools for trimming composite part edges and cutting cores. Typically, finer tooth blades do a better job. If you're in the market for a band saw, try to stick to conventional 2-wheel models. Anything rated for cutting wood works well on composites. Nine-inch models like the Delta Shop Master pictured above (the model I own) are low cost, available from box stores and online, and have a small footprint. Make sure to get up to speed in band saw operations and safety procedures. If in doubt, use a push-stick to push something through rather than your hands. As for blades, Powertec brand blades available on Amazon have served me well, and are available in multiple lengths and tooth TPI sizes.

2.12 - Rotary Tools

Commonly known by the tradename Dremel, rotary tools are the go-to wonder tool for composites. You can surface prep for a bond via grinding stones or sanding drums and cut parts with carbon cut off wheels, to name just a few uses.

If you need to scale up to more power, there is a tool called a Roto-Zip that is basically an up-scaled Dremel.

Warning! Avoid any kind of cutting tool that oscillates, like a jig saw or scroll saw, when working with laminated parts. These reciprocal-motion tools can lead to edge delamination in your laminated part if the blade catches in the material. Band saws and rotary tools, like those made by Dremel and Roto-Zip, always run the cutting blade in one direction, and are laminate friendly.

Chapter 3 - Raw Materials

3.1 - Reinforcements: Fiberglass, Carbon, and Kevlar

In this section, I will discuss the common composite materials I use on my own plane builds and other composite projects.

www.gouldaero.com/pmc also has links where you can buy many of these supplies.

Each kind of Fiberglass, Carbon, or Kevlar typically has its own designation number, similar to how metals like 2024-T3 or 4130N are designated by numbers/letters signifying their specific properties. Lamination systems also have their own numbers/names. With prepreg materials, typical nomenclature is "reinforcement/resin system" i.e. 7781/2510, which signifies Toray 7781 fiberglass and 2510 Prepreg resin. Long-established companies like Hexcel originally created some of these styles, and now other textile manufactures produce these to the same spec. Other companies may have their own proprietary materials; particularly in the case of carbon fiber.

For Fiberglass and Carbon dry goods I usually like:

AVT Composites: http://www.avtcomposites.com/

Fibre Glast: http://www.fibreglast.com/

Aircraft Spruce & Specialty (ACS) usually has higher prices on fiberglass cloth styles, and sometimes does not give the same packaging attention as AVT and Fibre Glast. In a pinch though, ACS can be another source of materials.

3.1.1 - Fiberglass

"The Rutan Cloths" 7725 and 7715

Much of the pioneering in DIY composite home-built construction was done by Burt Rutan and his Rutan Aircraft Factory "RAF" that developed and sold plans for aircraft like the Veri-ez, Long-ez, and Defiant. Through much R&D back in the 1970s, he found 7725 and 7715 fiberglass cloths to have the best balance of mechanical properties and ease of use; and they are still useful today.

7725 is a 2X2 twill useful for going around compound curves (cowlings, fairings, wing tips, etc.). Wet-out characteristics rank as medium. Among Long-ez, Cozy, Quicke, and Velocity builders, 7725 is usually referred to as "bid" for bi-directional. 7725, along with any other 2X2 twill, can become quite distorted if not handled carefully. So make sure to test it out before using it on something critical.

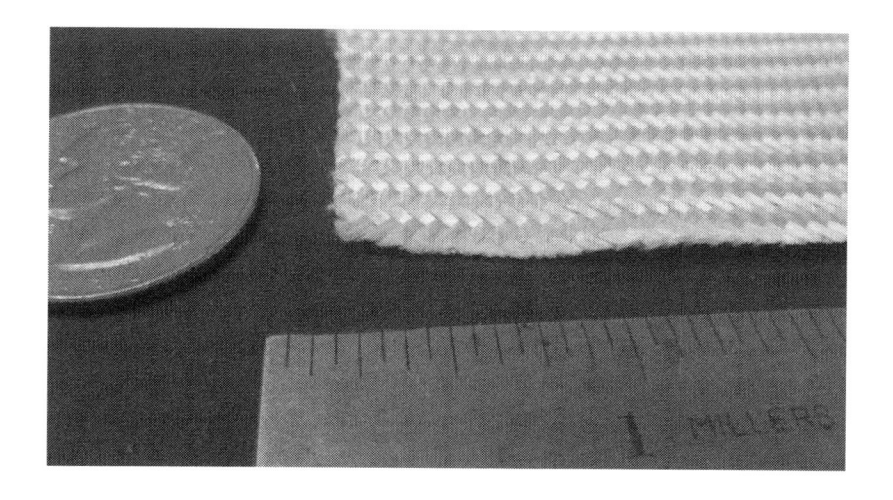

7715 is considered unidirectional in the home-built world, and is referred to as "Uni". Most of its strength goes in the "1", or off the roll, direction. 7715 is useful for the 90° plies for shear (if 0° is the long axis). 7715 can also be used for 0° longeron or spar cap plies in major or minor structures. In considering cost, weight, and effort, in constructing the main spar on the wing and stabilizer, carbon wins out for the bulk cap material. That being said, 7715 still has its use in smaller, less bulky layups. 7715 is also good to use for joint tapes where you need a dedicated ply orientated to go around the bend to take on the primary load path for joint separation loads.

3733 is another commonly used cloth. From a composite design, engineering, and fabrication perspective, 3733 is my favorite fiberglass cloth for general layups, outer skins, shear web, and torsion ply materials. The specs and testing place this cloth on the lean side of e-glass fiberglass products for strength to weight. The fabric pattern, depending on if it's twisted, flat, or high-density nesting flat can be a big player in strength-to-weight. 3733 is very easy to wet out and has excellent handling characteristics. It's also a lower weight cloth than 7725 or 7781 making for less of a jump for your structural need = lower part weight. This also gives you better ply count resolution allowing you to carry less dead weight along in high ply count / taped ply applications.

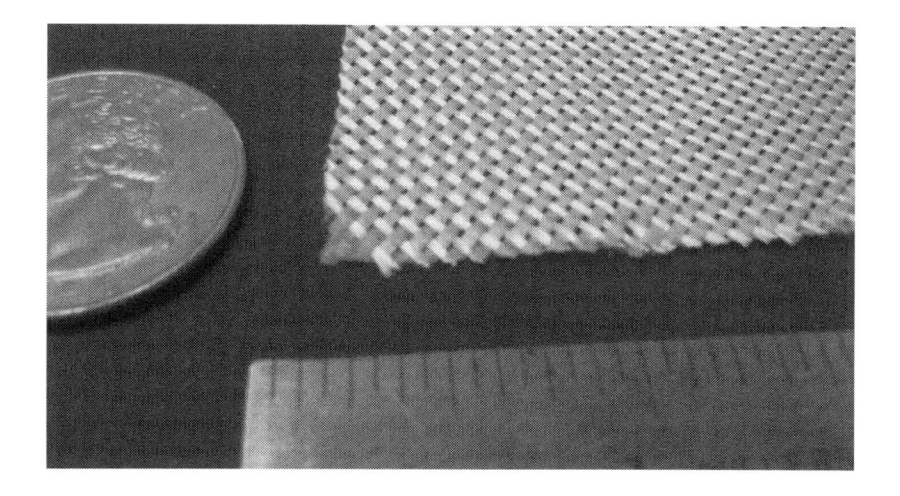

AGATE 7781/MGS418 Wet Layup Data (Source AGATE / NIAR)

3.1.1 Summary

MATERIAL:	Lancair 7781 Glass Fabric / MGS 418 Wet Layup		7781/MGS 418 Summary
FIBER:	BGF 7781 Glass Fabric	RESIN:	MGS 418
T_g (dry): 216.4°F	T_g (wet): 209.9°F	T_g METHOD:	DMA (SRM 18-94)
PROCESSING:	Vacuum bag cure (22+ in. Hg.): 170-230 ± 10°F for 5.5-24 hours		

Date of fiber manufacture	10/9/98	Date of testing	10/26/99 – 12/15/99
Date of resin manufacture	Unknown	Date of data submittal	12/15/99
Date of composite manufacture	6/98 – 3/99	Date of analysis	10/28/99 – 12/15/99

LAMINA MECHANICAL PROPERTY SUMMARY
Data Reported as: Measured
(Normalized by CPT=0.0108 in.)

		CTD		RTD		ETD		ETW	
		B-Basis	Mean	B-Basis	Mean	B-Basis	Mean	B-Basis	Mean
F_2^{tu}	(ksi)	52.02 (54.54)	60.89 (59.83)	41.85 (41.25)	48.99 (47.89)	36.06 (35.48)	42.21 (41.19)	31.05 (33.00)	36.34 (38.31)
E_2^t	(Msi)	---	3.42 (3.37)	---	2.39 (2.33)	---	2.10 (2.05)	---	2.65 (2.82)
v_{21}^{tu}	(ksi)	---	0.175	---	0.118	---	0.087	---	0.107
F_2^{cu}	(ksi)	58.13 (57.17)	66.72 (67.30)	45.11 (42.64)	51.77 (50.19)	27.41 (25.93)	31.45 (30.52)	25.35 (25.20)	29.10 (29.67)
E_2^c	(Msi)	---	2.27 (2.31)	---	3.16 (3.07)	---	3.06 (2.97)	---	2.14 (2.17)
F_{12}^{su}	(ksi)	19.67	21.16	14.36	15.45	8.74	9.41	7.68	8.26
G_{12}^s	(Msi)	---	0.63	---	0.44	---	0.23	---	0.25
F_{13}^{su**}	(ksi)	---	---	7.00	7.59	---	---	---	---

** *Apparent* interlaminar shear strength

7781 is "The Infamous Aerospace Cloth". Chances are if you see a certified or military fiberglass composite part, it's made with 7781. 7781 is a 8HS cloth configuration (under 7 over 1…). 7781 has a good balance of flat and high contour performance. Fabric resolution is in the 50 threads per inch range (very high), allowing the material to act more uniformly in smaller cross-sections. For example, edges with riveted nut plate and edge distance scenarios. However, wet layup is painful. Epoxy batches should be small, and used quickly for minimum working viscosity. Also, I highly recommend using a small short-nap "pink" roller to process the layup. Speed is key; you need fresh resin "in" the cloth, Quick!

6781 is the slightly stronger s-glass version of 7781. 6781 looks almost identical to 7781, however, due to the slight glass chemistry differences between e-glass and s-glass, 6781 has a slight multi-spectrum reflection visual effect. Cost wise, 6781 is in between 7781 and 282/284 carbon. However, unless you're boxed into a hard requirement for a "fiberglass material", I suggest jumping up to the two carbon options. 6781 was used as P2 @ 45° for the shear web portion of the flaps on my plane.

3.1.2 - Carbon Fiber

282 is a great plain-weave, basic-use carbon cloth. Both 282 and 284 carbon styles have become quite affordable. 282 is an excellent choice for the shear ply in flight control surfaces (elevator, rudder, ailerons). This application is where you will get the most bang for the buck with carbon. If your budget allows, carbon can be used in other major parts of your project.

284 is a 2X2 twill, and is a main "go-to" for a "decorative" carbon used on items like car hoods and motorcycle fenders. In fact, black 2X2 twill is usually called "carbon look". In a structural application, it's good for areas with compound curvature (curvature going in 2 different directions).

3.1.3 - Kevlar

Kevlar does have some strong, but limited, uses. Beyond the more well-known use for bullet proof vests, in a composite layup, it's good in high-energy-absorption areas; like near a landing gear or engine mount. However, Kevlar has some major drawbacks. Kevlar is very difficult to cut, even with high cost specialty "Kevlar-cutting" scissors. Kevlar has decent strength in tension, but is weak in compression; unlike carbon and fiberglass which are more balanced.

Also, I would like to give a major heads up to the boat guys reading this book. Kevlar absorbs water, thereby lowering its material properties. In a conventional composite layup, Kevlar should have at least a low-weight fiberglass ply between the Kevlar and the outside world. Additionally, using Kevlar as the last layer will result in an un-desirable surface finish if you sand the composite surface. Any kind of sanding will result in the surface "pilling", or fuzz balling up; and more sanding will only make a bigger mess.

3.1.4 - Pultruded Materials

Carbon fiber and fiberglass can be purchased in pre-made "pultruded" rods in a variety of cross sections. Pultrusion is similar to an extrusion, except the material is pulled through a forming die rather than pushed. In the Aerospace world, much pioneering development work with pultruded carbon was done for the V-22 Osprey rotor blades. Marske Aircraft http://www.marskeaircraft.com/carbon-rod.html is one of the main suppliers of this material. Pultruded material can make quick and precise work of making spar caps. Pictured below is the material I used on my Aircraft Design. It comes supplied in a 3-4' diameter roll.

Some items to note about working with this material:

- Cut it with a rotary tool or band saw (using snips results in split ends).
- The material must be de-flashed prior to use. With square stock, a clamped-up setup and a random orbit sander does the job. With round stock, chucking it in a drill or lathe with sand paper would do the trick. The goal is to get rid of the "shiny black" surface finish, and produce a dull black finish. If left with a shiny surface finish, the smooth surface makes an excellent release agent, and your resin will not adhere.

Prior to using the material, it should be vacuumed and cleaned off with isopropyl alcohol wipes.

You must vacuum bag or clamp down the material for cure since it comes in a spool and has some amount of "memory". Without proper clamping pressure, the material will be wavy on a layup table. You will need to hold it down during cure to get a viable/useable product.

3.1.5 - Prepreg Materials

Over the last 2 decades, out-of-autoclave cured Prepreg materials have gained industry acceptance in part 23 certified aircraft. The big push behind this effort is the AGATE program (Advanced General Aviation Transport Experiments) that was co-chaired by the FAA, DoD, and NASA, and facilitated through Wichita State University. NCAMP (National center for Advanced Materials Performance) is the follow-on program to AGATE. Note: The AGATE data is still used on projects today and offers the best turnkey solution to make certified Aircraft Parts. This was the key reason for using prepreg as the prime material on the NASA wind tunnel parts. It's as turnkey as you can get when it comes to composites.

The AGATE web page and database can be found at:

http://www.niar.wichita.edu/agate/

The benefit of prepreg is it takes much of the effort out of the layup and design process, since you're getting a known and uniform resin content. If you are in the certified aviation industry, prepreg will be your most turnkey option for a FAA certified part. Additionally, you get a decent amount of working time with Prepreg.

However, there are some challenges to using prepreg. One of the biggest is facility cost. At minimum, you will need a dedicated large storage freezer with a backup power supply to store the Prepreg materials. Depending on the application, you may need a large, stringent clean room. Minimally, you will need an oven large enough and hot enough with a vacuum pump system. If you need an autoclave for your work, this can add significant cost. As your part size and/or count increase, so too will the need for prepreg processing equipment.

Toray Prepregs:

7781/#2510 fiberglass prepreg manufactured by Toray Composites, and being the major "yards sold/used in certified airframes", is the AGATE winner. Cirrus, Columbia, and Liberty airframes are made out of the stuff.

T700SC-12K-50C/#2510 carbon fiber cloth prepreg is what makes up the bulk of ultra-high-performance airframes like Lancair.

T700SC-12K-31E/#2510 carbon fiber uni prepreg is a good choice in spar caps and longerons.

3.1.6 - Peel Ply

Peel Ply has several uses in composite construction. It can be used on the mold surface or mold table surface to provide a level of protection from release agents and the laminated part. It can be used on areas that will be bonded or receive additional laminations later on. Also, it can be used on top of an additionally laminated part's edges to keep any potential lift-up from happening, while providing a better cosmetic appearance.

3.1.7 - Release Film

Release films are typically used in prepreg fabrication. They are usually perforated, allowing air to escape through the laminate into the breather system and thereby prevent the breather from becoming stuck onto the laminate. Peel Ply can also serve the function of release film.

3.1.8 - Vacuum Bagging Film

Vacuum bagging film can be highly specialized and high-cost. Bagging film is typically used with BMI/autoclave laminates, but can also be used with hardware store plastic and room temp laminates in the garage. If you're using prepreg, you will need high temperature bagging film as some of the lower cost options can lose vacuum under the heat.

3.1.9 - Transfer Plastic

Transfer plastic is one of the main secrets for achieving quick, professional results in the shop. This material is simply party table plastic that can be purchased at stores like Party City, Michaels Crafts, and even online. For its cost, and what it does, your money will go a long way. This material does not permanently adhere to epoxy and is easy to remove once a ply is in position and in contact. When using this method, you are basically creating your own on-demand prepreg that cures at room temperature. You can use transfer plastic to maintain fiber orientation and a precise edge of a ply. With carbon laminates, you can even see the back side of the laminate to ensure you have no dry spots.

Transfer plastic is available in multiple colors, but I recommend green, orange, and red as it's a great visual mistake-proofing to not leave it behind in the laminate. I will have my dry goods cut oversized, and trim them with my dedicated wet scissors once wetted out and squeegeed. You can also use a rotary razor "pizza cutter", but you will want to have a dedicated cutting table for this and not your laminating table. Also, you will need to clean this style of cutter right away to keep it from locking up after cure.

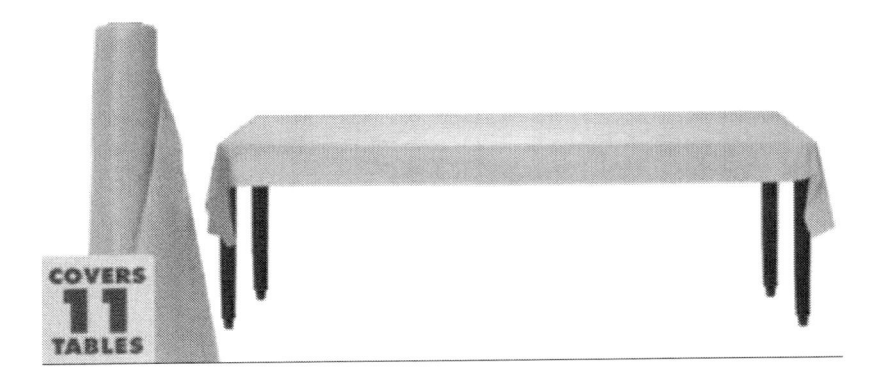

3.2 – Resin Systems

3.2.1 - Epoxy Resin Systems

Aeropoxy PH 3660 is the main lamination system I used on my plane. Coupon testing shows it to be on par with other systems. PH 3660 can be purchased direct from Aeropoxy, AVT, or Aircraft Spruce. Usually they bring some to Sun-N-Fun or Oshkosh for sale. PH 3660 has a long shelf life, and unless you are going through gallons a month, should be purchased in gallon kits instead of 5 or 55-gallon sizes.

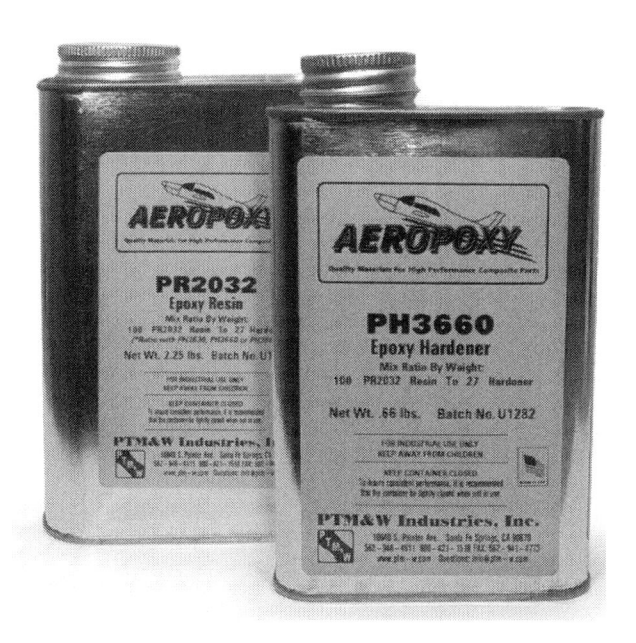

MGS is the other good off-the-shelf aircraft-grade lamination system. MGS lamination systems are listed in the AGATE wet layup allowables PDFs.

West System epoxy is another option, and some people insist on building airplanes out of it. However, it's more common in the boat and automotive realms. It is one of the most available epoxies out there to get at the local brick and mortar store. You want to go with the longer-curing hardeners (ref 206), as the West System's quick-curing (ref 205), like the name suggests, gives you very little time to get everything together. West System offers a low cost "ketchup" pump system that works quite well.

3.2.2 - Vinyl Ester Resins

Vinyl ester is an in-between product (Epoxy vs. Polyester). It is highly common in the marine industry, providing enhanced corrosion resistance and lower heat distortion. Glasair made extensive use of vinyl ester resins in their product line.

3.2.3 - Polyester Resins

Polyester is the lowest cost resin option. The fumes can be quite extreme, and it can melt some foams like Styrofoam on contact. Its pot life is controlled by the ratio of resin to hardening agent. Unlike epoxy, the hardener for polyester is applied as only a few drops. This hardener is sometimes referred to as "catalyst".

3.3 - Cores

3.3.1 - Foam Cores

Foam cores fall into two format types: Sheets to be used as is (like a Glasair), and blocks that can be cut to a desired shape.

Sheet foams come in several different chemical makeups.

Divinycell is a common PVC-based foam core, and is one of the main go-to sheet foams. It is inert to fuel and is used on items like a Long-ez fuselage and fuel wing strakes. It's a closed cell foam typically used in the 3 (H45) - 6 (H100) Lbs/Ft^3 density range. It has high compression strength, durability, and excellent fire resistance. Divinycell can be vacuum formed to compound shapes, and can be bent using heat. Divinycell is compatible with the common resin systems.

Last-A-Foam is a fine, closed-cell polyurethane made by general plastics. Its main use is in composite tooling where it's typically CNC routed into a desired shape. It can also be used as a core material on aircraft and other vehicles. After machining, the surface should be completely vacuumed using a shop vacuum with a brush attachment just before lamination. There has been some "internet buzz" about this product being undesirable as an aircraft core; however, Last-A-Foam is considerably tougher and less friable than a pure urethane core. The pure urethane core is used in the KR series of aircraft that has been flying since the 70s (with some airframes logging over 1000 hours), all with an exposed internal core in the wings. It may not be ideal, but there is a proven airworthy track record.

General urethane foam, as mentioned above is the go-to material on KR's. It is an older chemistry and is very friable. It can serve well as a tooling material where you need to hand carve something into shape, like a cowling inlet, ergonomic seat, or console items in the cockpit.

The other foam core material available is Styrofoam. Styrofoam is commonly available in blocks that can be hot-wired into shape (like a Long-ez wing and canard).

An important characteristic to note with Styrofoam is you can't expose it to gasoline. Gasoline will dissolve Styrofoam! The Long-ez's hot-wire-cut Styrofoam structures all bolt onto the fuselage, and do not have a fuel leakage conduit path to the Styrofoam cores. It is worth noting again that the sheet foams on the market come in several chemistries, and most are inert to gasoline and oil.

With all foam cores, a medium viscosity (cream peanut butter) consistency mix of Micro (Epoxy and Micro Balloons) should be applied to the surface and squeegeed down to the contour. If the Micro mix is not going into the pours, and is balling up on the surface, you will need to add more epoxy to the mix until it stays within the open pores of the foam core.

3.3.2 - Honeycomb Core

The most common honeycomb used in aircraft and car construction is Nomex. Vacuum bagging or compression molding is needed to ensure the face sheets make full contact with the edges of the honeycomb walls. They will also need to have a slightly resin-rich exposed side of the face sheet that makes contact with the honeycomb for wet layups, or will need adhesive film for prepreg layups. If it's your first go-through with honeycomb, you should test your methods and setup on a 2-3 ply count fiberglass/Nomex sandwich to visually ensure that your core adhesion is correct. The cell configuration is another key driver to look at with honeycomb. The default conventional configuration is nested Hexagons.

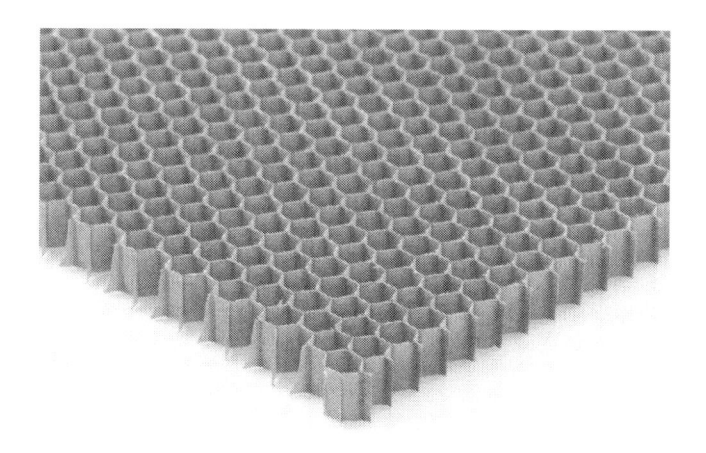

This is good for most applications with moderate curvature. There is a drawback, however, with a relatively thick core being used in a tight bend radius (like trying to wrap it around a small diameter cylinder). In this situation, hexagon core will form a saddle shape, where the other axis will want to lift off on the edges. This can produce significant fabrication limitations. Without better options, a thinner core or relief cuts must be used.

The better fix is to use an over expanded (OX) core that allows for tighter contours in one direction.

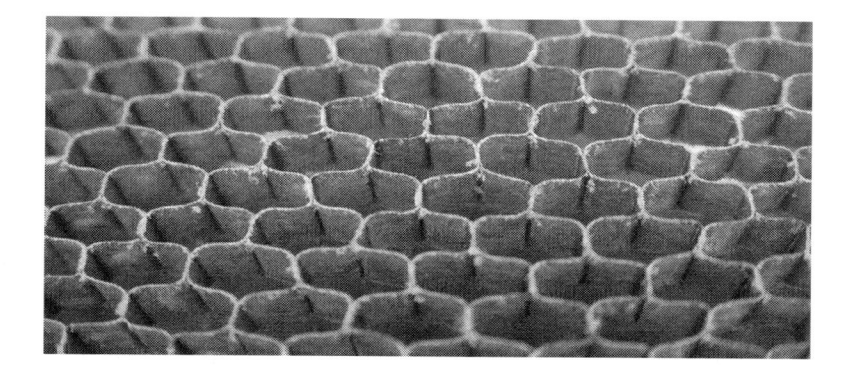

Note: if you are trying to bend honeycomb around two tight radii directions, such as wrapping it around a ball shape, you will likely need to divide the honeycomb core material into strips (or dart it) to keep it from fighting you during lamination.

3.3.3 - Wood Cores

One of the early adapters of what we would now think of as sandwich aircraft construction was the De Havilland Mosquito, which used Balsa cores and Birch plywood skins.

Balsa is still used today in applications like the Chevy Corvette floorboards.

Balsa cores are a greener option when compared to Nomex, aluminum, urethane, and other polymer cores. Balsa trees grow rapidly and are harvested in less than a decade. When using Balsa cores, attention should be paid to the effects of humidity and water vs. desired mechanical properties and long-term durability (primarily with Boats).

3.4 - Inserts & Hard Points

Almost every composite structure used on an aircraft, car, or boat, will need to be mechanically fastened to some other component. While composites are great at distributed loads, the advantages degrade with point source loads in a uniform composite structure. This is where hard points come into play. Hard points are designed to pick up the point source load and distribute it to the composite structure.

Optimized hard points have two key features: The first feature is a core replacement, where the local area has the traditional lower structural capability core material (like honeycomb or foam) replaced with a material of higher structural capability. These inserts are added at the same time as the core, and the face sheets cover both sides of the insert, as they would with a core. There may be some plies that extend past the insert 2-3 inches beyond the insert border to reduce the jump in structural capacity (spread it out). The second feature is to have a metal bushing passing through the structure or external plating to the outside of the structure. This bushing serves to define the transition between distributed loads and point source loads.

3.4.1 - G10-FR4

From my composite experience, G10 is the ideal core substitute hard point material. G10 is a pre-made fiberglass/epoxy sheet that is press-molded, and has properties in the same ballpark as autoclave-cured prepreg fiberglass/epoxy. Since G10 is fiberglass and epoxy, it will not galvanically react with metals. The fiberglass G10 is made with is an ultra-high-resolution "thread count" material, thus making the matrix ideal for point source load transfer. Also, in a carbon face sheet application, since it would be a carbon - G10 - carbon sandwich setup for bending considerations (sig = Mc/I), all the high stress is in the "outer most fiber" – all the way out in the carbon, while the G10 is in the lower-stress cross-section centroid area.

From a cost perspective on a volume basis, G10 is equal to the cost of just the raw fiberglass and epoxy. You would still need a high-power press or autoclave to even attempt to duplicate this pre-manufactured material; not to mention the manual or even robotic effort required to fabricate the sheet. In summary if you have a large flat sheet to layup or insert, cost, time, and strength wise, you will come out far ahead using G10 over a layup.

G10 can be purchased from composite specialty shops or McMaster Carr. The surfaces are slick, so you will need to rough them up with a dual action "DA" sander prior to layup. McMaster Carr offers colored G10 that aids in the surface prep process because it's easier to visually see what is "dull" and done versus "shiny" and still requiring sanding. G10 can be cut with all the conventional cutting methods; however, band saws are the best option.

The shear webs and wing connection hard points on my center spar cox are made from G-10. The caps are graphlite pultruded carbon. It's as efficient as you can get.

3.4.2 - Phenolic

Phenolic is one of the oldest modern composite materials. Bakelite was one of the first commercial uses for phenolic. Phenolic is more heat resistant than G10. It is also a common material used in aircraft control system pulleys due to its wear characteristics. The material properties of pre-made phenolic sheet are much lower than G10. Like G10, it can be cut with conventional composite cutting methods. The exposed faces are shiny and must be prepped for proper lamination adhesion by sanding. Note: Phenolic lets out a very strong odor when cut and sanded. Also, a phenolic sheet can warp over time, so it's best to keep it constrained flat (like storing plywood). Unless your specific design demands phenolic, or you're going for the look of it, G10 is a far better choice.

3.4.3 - Carbon Sheet

Pre-made carbon sheet material has gained great popularity with racing drones, "multi rotors", RC helicopters, RC cars, and for its visual appeal.

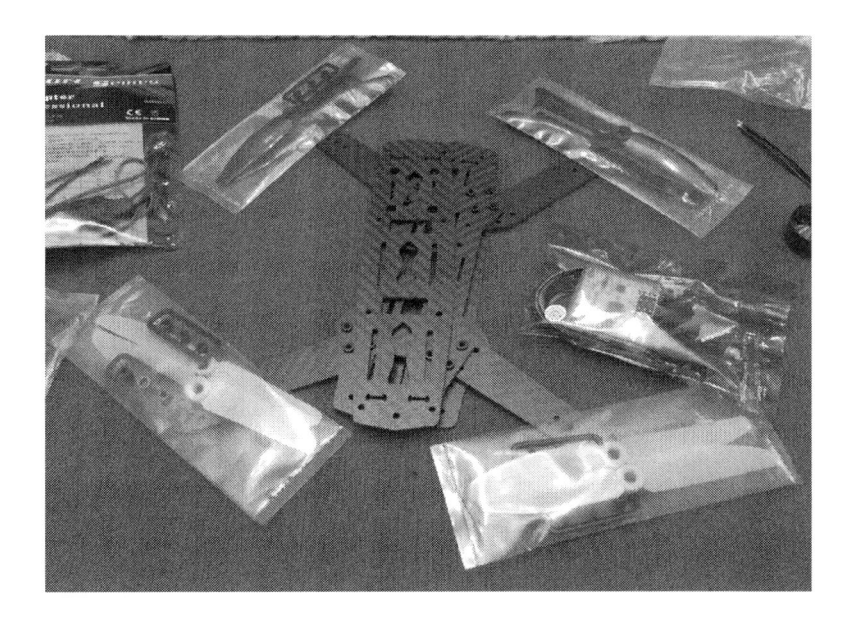

However, carbon sheet is not as ideal of a hard point insert material as G10. Most pre-made carbon sheet is in the 10-20 threads per inch (TPI) resolution range. When compared to G10's 30 - 80 TPI range, you can see it makes for a less ideal load transfer to a point source load. Being made from carbon, carbon sheet must be insulated from metal objects passing through it to avoid galvanic action.

3.4.4 - Wood

Wood has been a go-to insert material in aircraft and cars. However, it has some drawbacks. To get strength properties in two directions, like you would from G-10, phenolic, and carbon sheet, you need to use high-ply-count plywood. Moisture absorption and general rot can be an issue, particularly with boats. One of my co-workers was given a hand-me-down boat (fiberglass over wood) that was in need of some "TLC". It turned out that the transom and stringers (made of wood) where all rotten. In the end, he likely would have been faster and achieved a better result using the boat hull as a mold and starting anew.

3.4.5 - Aluminum

Aluminum can make an excellent bushing and plaiting option for a composite structure. Aluminum is corrosion resistant, easy to work with, and is well-proven in the aerospace industry. However, I would recommend avoiding Aluminum as a core substitute material unless you need to directly tap something through a composite structure, like a NPT thread for a fuel pickup.

3.4.6 - Steel

Steel can also be used as a hard point, but some additional challenges arise over aluminum. Steel is typically harder to cut, finish, and drill than aluminum. Steel is also more prone to corrosion. The chrome alloying element in stainless steels makes an excellent release agent for composite resin systems, and generally should be avoided.

3.4.7 - General Metal Prep for Inserts, Bushings & Jacketing

The parts of the metal that will be in bonded contact with the composite part need to be roughed-up just before bonding/laminating. Aluminum forms a protective oxide layer very quickly after being sanded, so consider any prep of an aluminum component to be part of the active "hear and now" layup or bonding sequence. For bushings and jacketing, I would recommend an epoxy-based adhesive intended for metal bonding, like Hysol or PTM&W ES 6228. For jacketing (metal panels and components bonded to the outside of a layup), I would recommend an internal insert core of G10; and to rivet through the part for not only a chemical bond, but also a mechanical bond.

3.5 - Fillers

Fillers are added to lamination and bonding products to increase viscosity, material properties, and/or reduce the density. Almost all fillers "aerate" extremely easily, particularly when poured. Typically, a scoop is the best way to get the filler into your resin mix. You should wear your respirator when handling any of the fillers, even if it's to transfer the material from one container to another. In the beginning, they all kind of look the same, so it's good to keep a label on them. With more exposure and experience, you can start to tell them apart, even with a small sample.

General note for mixing epoxy with the fillers: You want to get the consistency just "wet enough" to stick, if it goes on dry, the material crumbles off and it will do no good. At the same time, you don't want to get the mix so wet that is runs quickly.

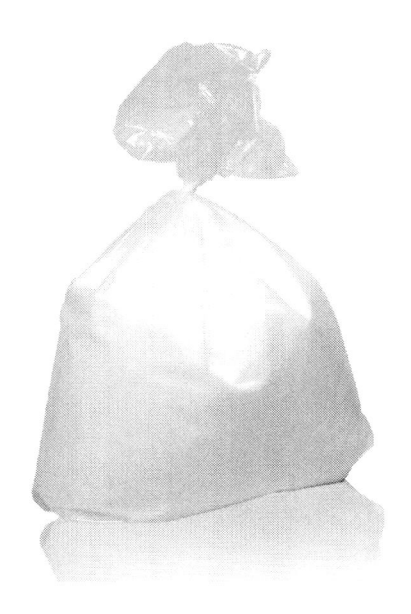

3.5.1 - Micro Balloons

Micro Balloons (sometimes called Micro Spheres or Glass Bubbles) are hollow balls of glass that are extremely lightweight. For wet layup, micro balloons are the prime thickening agent added to lamination epoxy for hard surfacing a foam core. Micro balloons are also the go-to finishing filler material used on outside cosmetic surfaces. Micro balloons are one of the lightest filler materials available. Peanut butter consistency is the common viscosity to shoot for with this mix. You can increase or decrease the viscosity of the mix by varying the ratio of lamination epoxy to micro balloons.

3.5.2 - Flox

Flox (sometimes called Flocked Cotton Fiber) is a by-product of cotton production. Its prime use in composites is for structural corner fillets that will receive a connection laminate over the corner. Flox offers better material properties in comparison to micro balloons, however, to improve its spreadability, micro balloons (in small amounts) are sometimes added to the mix.

3.5.3 - Milled Glass Fibers

This material is made by milling fiberglass into a very thin consistency. It can be used as a structural filler, but can become heavy. Flox is more of the go-to for this application.

3.5.4 - Cab-o-sil

Cab-o-sil is a fumed lightweight silica thickener. It's tougher than micro balloons, however it is heavier and harder to sand.

3.6 - Adhesives

At some point in your project, you will need to structurally adhere one part to another. For this, you will need adhesives.

3.6.1 - Epoxy Adhesives

Epoxy adhesives are your main go-to for structural adhesives. They come in many types and grades, from home and hobby Devcon epoxy sold in the box stores, to aerospace-grade Hysol. For most composite work, I recommend two products. Both are made by PTM&W industries. If your budget or requirements aim for the next notch up, I would look at Hysol's product line.

ES6209 is an excellent composite-to-composite adhesive. It's a yellow/straw color when mixed and cured. The mix is a simple 1:1 ratio.

ES6228 is an excellent composite-to-metal adhesive. It is a grey color when mixed and cured. The mix is also a simple 1:1 ratio.

There are three different ways of mixing ES6228:
- Pre-marked and calibrated heights on 9oz disposable plastic cups from Sam's Club.
- Medicine cups are usually my go-to for average jobs.
- 10ml Syringes for Small Jobs.

3.6.2 - Cyanoacrylate

Cyanoacrylate is commonly known as "CA", instant glue, or model-airplane glue. This is the go-to "tacking" agent (similar to tack welding on 4130N) for composite structures. It dries quickly and is safe to leave behind while more substantial methods like epoxy and/or Flox joints take over. The best general use "CA" brand I have found is made by Bob Smith Industries. www.bsi-inc.com

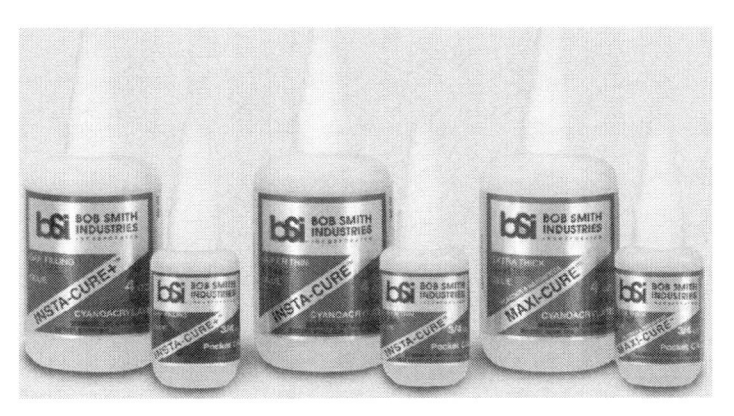

Like with all adhesives and other chemicals, always wear your safety glasses. Always store CA away from children. Never store CA in your medicine cabinet, as it could be confused as an ingestible or eye drops. CA can produce strong fumes when it is curing, and the fumes typically go

straight up. It is best if you stand to the side of your part, not directly above it.

3.6.3 - Polyurethanes

One of the common commercially available polyurethane glues is Gorilla Glue. It expands as it cures, and its prime function in composite construction is joining sections of foam cores or honeycomb cores. In addition, polyurethane glue can be an alternative to micro balloons/epoxy joining of cores.

Chapter 4 - Gigs, Fixtures, Molds, & Moldless Methods

4.1 - The Value of Tape and Single-Use Foam Molds

Duct Tape is the go-to material for producing various types of composite layups. The back side of duct tape does not adhere to epoxy, so you can wrap foam with duct tape and use it as a quick mold. Here is a quick duct tape "mold" on a C5 Corvette engine bay part being used to make a contour-conforming patch panel.

Here is the finished part ready for bonding. As you can see, this quick and easy duct tape mold allowed this part to pick up every contour in this area with minimal effort.

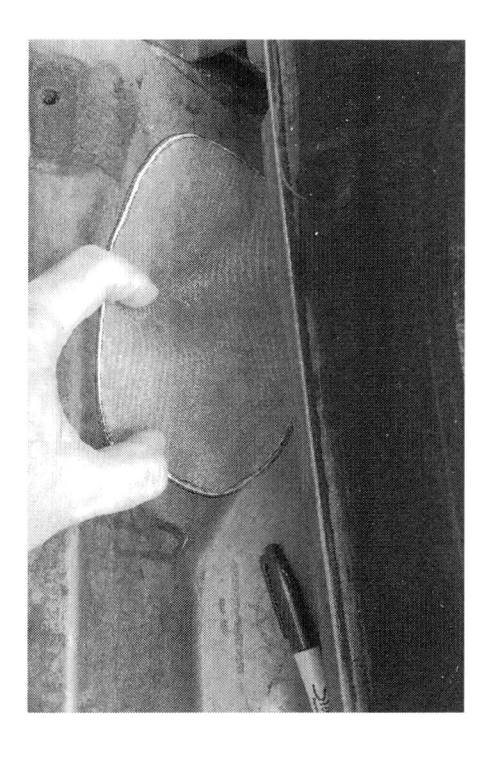

I've used the same method to make exhaust fairings for Reno race aircraft by putting duct tape over the contoured end of a #80 Argon cylinder from a TIG welder. Start looking around and you will see all kinds of things you can use for molds to make parts for your project.

When coupled with foam, duct tape molds can have extreme negative draft on them. You can hog out the foam with a Dremel and then separate the duct tape from the part like a liner. Here is a control system carbon/fiberglass/G10 bracket that was molded by sticking duct tape to an MDF board and 1 lbs/Ft^3 polystyrene foam sides (also lined with duct tape). The foam was hot glued to the MDF, and was a one-time-use mold.

Word of caution with Duct Tape: Only use it for room temperature laminating/bonding applicaExtreme heat like in prepreg processing will make a mess.

4.2 - MDF

MDF is an excellent mold making medium for reusable molds. The fine structure of the material allows for intricate contour work with minimal backfill.

In this example, I used four blocks of MDF and two retaining rings to make the ribs on my plane. It's a rectangular plan-form, so I was able to use this same mold around two-dozen times. To remove the part, the eight bolts and retaining rings are removed, and the four blocks all pop off one at a time.

4.3 - Partial Molding Techniques

Partial molding allows you to transition from a core to a pure laminate section with almost no re-work, aside from secondary lamination surface preparation. Here is a fuselage bottom skin that has an access door setup in the cores. The foam cores are nailed with 4-penny flush head nails to a 1/8" thick hardboard table that is supported with a 1"x4" frame. The hardboard is covered with duct tape or packing tape with release film. Peel ply is taped to the table beyond the borders of the pure laminate area to aid in the overall layup. 4-penny flush heads are sufficient enough to hold down the core, but small enough to be pulled through the core when you release it from the molding table.

The part is laid up with the fiberglass transitioning from the core to the table, and back up to another section of core. Peel ply is added around the area because additional operations will be done in this area at later time.

With the part pulled from the partial molding table, it is almost ready for a layup on the other side.

In a pinch, or on smaller parts, duct tape over polystyrene foam can serve as a partial molding table because the nails can easily pass through both the core and the foam block.

4.4 - Vacuum Bagging

You can use vacuum bagging at home. The Cozy girls demonstrate a method at Sun-N-Fun using a fish aquarium pump and Stretch-Tite for small composite parts with no tape or bonding clay; just the static cling of the Stretch-Tite holds a seal. The delta P with this method is low, but can still produce satisfactory results.

I use a middle ground approach:
- Visqueen plastic from the home improvement store
- Blue tape for seal
- Peel Ply as the release film/ply
- Paper towels as the breathing media
- Transfer plastic on top of this
- Cloth store felt near the pickups
- An old, but decent sized, vacuum pump.

Here is this setup in action with my Graphlite pultruded carbon spars for my plane, vacuum bagging is necessary to clamp down the material during lamination into a spar cap form to counteract the slight but natural wave that the material has in its raw form.

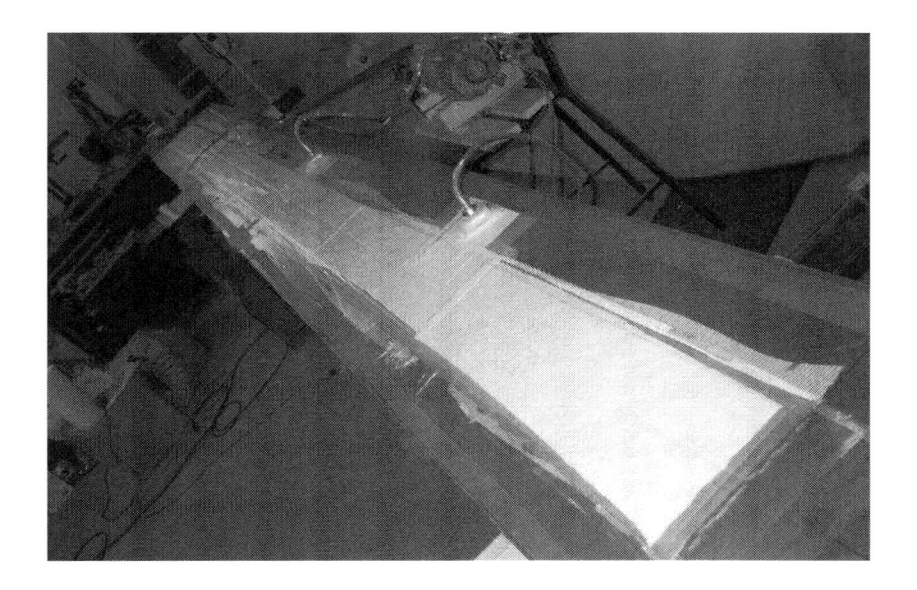

As for vacuum bagging, the sky is the limit. You can buy expensive high-temp vacuum bagging film, media, tape, and sealing compound, and even an autoclave to ramp up the pressure differential.

4.5 - Hot Wire Foam Cores

You can make your own hot wire setup. With some simple testing, you can achieve impressive results at home. It's typical to use doorbell transformers to step down the power, in conjunction with a rheostat/lightbulb dimmer switch to dial in the power correctly. It's best to make the frame out of wood (poplar or fine-grain pine), with a tensioning mechanism on one side and the hot wire on the other side. Use nickel wire from McMaster Carr for the actual hot wire for the best durability against the heat load. Typical hotwire thicknesses are .010" - .040".

Hardboard from the home improvement stores makes for an excellent hot wire template. The material can withstand the temperature and can be attached to the foam core with drywall screws.

There are services that will hot wire cut foam for you using CNC equipment, they can do custom ruled surface profiles or provide core kits for plans built aircraft like Long-ez's. Eureka CNC is one of the main services for this kind of work. http://www.eurekacnc.com/ Their prices are very competitive considering they are sourcing the foam, cutting it and shipping it to you.

Chapter 5 - Cloth Kits

You will need to have all component materials and necessary items ready before you laminate. You do not want to get into the habit of being half-way through a lamination, and realizing, "oops I forgot a...."

5.1 - Layout

If your laminated shapes are general rectangles, then a simple width/length list will do. Larger structures, like airplane wings and full car bodies, may be challenging to layup using the transfer plastic method because of their size. It is advisable to have some helpers, or to have fixtures/frames to hold the layup as you rotate and position plies over the part if you use a transfer plastic method.

The more manageable alternative is back-rolling where the fabric goes down dry on the part and is fully laminated in place. If you're doing resin infusion, the cloth goes down dry anyway, if you're doing prepreg, you simply leave the backing plastic on until the ply is in place, and then you can remove the plastic from both sides.

If you have a series of joint tapes to do, all with a common ply sequence and direction, you can lay them all out on the transfer plastic like so:

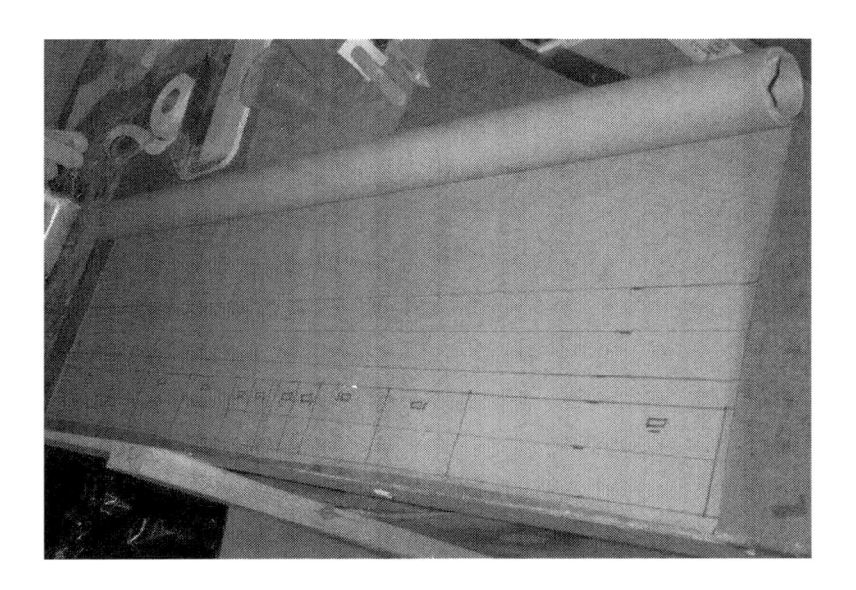

5.2 - Cloth Cutting: Methods Precautions and Considerations

You should always wear gloves when handling cloth. The oils and other contaminates from your hands act as a release agent. Cotton gloves work well for handling dry goods.

Dry fiberglass and carbon cloth can be cut with a rotary razor pizza cutter against hardboard or a dedicated cutting board.

Try to maintain ½" extra all around for the cloth compared to your final trim line.

5.3 - Kitting

You want all your plies, transfer plastic patterns, etc. ready to go before you laminate. Here is a set of back rolled kits ready for lamination on a spar.

5.4 - Back Rolling

For larger layups, like a wing, you will need to unroll the fabric and cut it to the correct size. Then, using an extra cardboard tube, roll the fabric back onto the tube; taking note of any directional sensitivity, like a trapezoid shaped object. Back rolling keeps the fibers aligned which prevents and/or reduces distortion, and allows for controlled positioning onto the surface that you are going to laminate.

Chapter 6 - The Game Plan

Doing a wet lay-up is a lot like pulling the pin and throwing a grenade. I'll never forget the comment a structures colleague made when I was at P&W: "ready-fire-aim". Before you start pouring out the part A and B, make sure the following has been accomplished:

6.1 - The Checklist

For a new core, make sure its profile has been cut, edges beveled and/or contoured (if applicable), and vacuumed for all loose dust.

If using Rutan style mold-less construction methods (cloth over foam / the foam is your alignment), you need to spend extra time making sure the core is exactly as you intend it to be. Doing some sanding on a foam core to get it into contour will take far less time and far-far less weight than trying to correct it after everything is laminated. Making up for contour mistakes on a laminated structure with copious amounts of micro balloons is an Epic Fail! This trick is only valid on molds and tooling that never see the sky!

6.2 - The Working Environment

Setting up the correct working environment is essential for successful composites work. Garages, hangars, and general workshops are the typical locations. They should be fairly clean and organized. In extreme climates, they should be air conditioned or heated as needed most of the time. Proper ventilation is needed during lamination. However, ventilation and can work against the climate control for a short duration while the lamination is happening. Once finished, you can start up the climate control again.

6.3 - Preempting, Eliminating, and Destroying Distractions!

"Human life critical" composites work (airplanes, cars, and boats) is a totally different animal compared to welding, sheet metal fab, and mechanically-connected structures that can accept work stoppages and restarts. It takes a level of focus, dedication, and completion that is rare to find in today's world of gold fish attention spans.

You are carrying out a process that involves chemicals that change from a liquid state to a solid state in a very short time span. **Remember: Once you poured the resin and hardener together, the pin on the grenade has been pulled. It's what you do next that counts!**

Preempt distractions by training your family and friends. You need to let them know what you are doing and share how any form of distraction for a set amount of time is an ABOSLUTE NO, baring a major medical emergency. In the modern era, where everyone is supposed to be on call to anyone without question, this is seen as a very rare request and outside of the norm.

One tactic to eliminate distractions is to choose days and times where you are not needed by work and family. You may need to put your work times into a calendar program like outlook. For people you are in regular contact with, including parents, spouse, and kids, it's best to send out a text that says, "I will be doing layup/composites/bonding... work from 6:00pm to 8:00pm and will not be near the phone."

Tim Ferriss is a personal development author (*4-Hour Work Week* and *Tools of Titans*) and one of my mentors. He is an absolute master of destroying distractions. His work is highly applicable to composite lamination and bonding where focus and attention to detail are paramount!

Beyond the "training" for those who are around you, it's best to put the phone on silent. Also, epoxy "gooie" on the gloves and an ultra-expensive smart phone screen is a bad combo!

One catch-22 that can come up is you should vent your workspace when working with laminating/bonding agents, and most people just open a garage door. This can invite unwanted distractions from passersby. By the time you satisfy the bi-standers' demanding questions, mistakes could have, or could be, made. "Was I on ply 2 or ply 22?" The best option is a dedicated wall-mounted attic fan with exterior vane that closes when not in use. This only requires a simple flip of a switch, and it's go time. The next best option is a box fan in a window. If these options will not work, you can resort to a slightly opened garage door with a box fan.

Turn on the music; it can help you focus. Music comes in three forms: radio, Pandora/I Heart Radio, and YouTube music playlists. There are multiple studies showing how music (in almost any form) can improve focus, productivity, and general Gung-Ho! Avoid TV/Cable and movies, or anything where you would normally pay primary attention to it or want to flip channels.

All cloth kits for the lay-up are cut and in front of you kitted up and ready to go.

6.4 - Composite Drawings & Layup Instructions.

On some projects, you may be designing your own stuff from scratch. On others, you may be following plans, instructions, or documents.

ASME Y14.37 has some basic coverage for composite drawings.

Here are the basics you need to know:

Ply Orientation Vector: Tells you on the print what way the ply is supposed to be going on the part. A "1" identification is the direction of the cloth coming off the roll, "2" is 90° to 1, or parallel to the roll spool. "3" is through the thickness direction.

Ply Sequence: Lets you know the order that the plies will go into the laminate. Typically follow a P1, P2, P3... format. The sequence is tied to the intended construction method as a methodized format. On a Rutan style (building from the core outward) construction, P1 (Ply one) would be the first ply to make contact with the core. Subsequent plies stack up from there. When laying up a part on a mold, P1 would be the first ply to make contact with the mold. Note: The "P #" could be fiberglass, carbon, and/or Kevlar for a given layup. Combining materials is particularly common with mostly carbon parts, where fiberglass is used as the ply closest to the surface; or with a galvanic insulating ply where a metal part will be making contact.

In prepreg layups, you will often be dealing with surfacing films. These are typically denoted as a (SF_) nomenclature. Additional nomenclature includes adhesive films (AF_) and cores (C_).

Edge of Ply: Tapering back plies is a very common method in composite construction, and is responsible for one of the many weight and material reduction benefits of composite construction when compared to metal. As the structural need reduces, so can the ply count. A laminated spar is a prime example where you are at maximum ply count at the connection, and reduce the count gradually as you get to the tip where the ply count is the lowest.

Overlap/No Overlap Zones: Most layups will allow you to overlap the plies based on the part size vs. the roll width, or the practical limitations of how big a ply can be. Laminated structures work on surface area, so the overlap width is critical for load transfer. The one instance you would not want overlap, and instead use a butt joint, is a molded part that on its opposite of mold side will bond to the joggled mold side of another part on a contour sensitive part.

Chapter 7 - Bonding

7.1 - General Alignment

This is where a great amount of attention should be paid. Once something is bonded, it can be considered "cast in stone". Digital levels and tape measures are the tools of the trade.

Here, a digital level is used to set the incidence of a horizontal stabilizer before the aluminum "L" clip brackets are line-drilled between the stabilizer and fuselage. Wedges and other clamping methods are used to ensure the two parts stay aligned.

Here, a digital level is used to zero-out the roll axis while setting a spar box. Note the wood wedges and Cleco/Aluminum L-brackets employed in alignment. Eventually, the spar box will be laminated to the fuselage sides and bottom, and the temporary Clecos, L-brackets, and wedges will be removed.

In this setup, a tail post block is temporarily screwed into the tail fuselage structure along with a control point/pivot screw in the middle of the block.

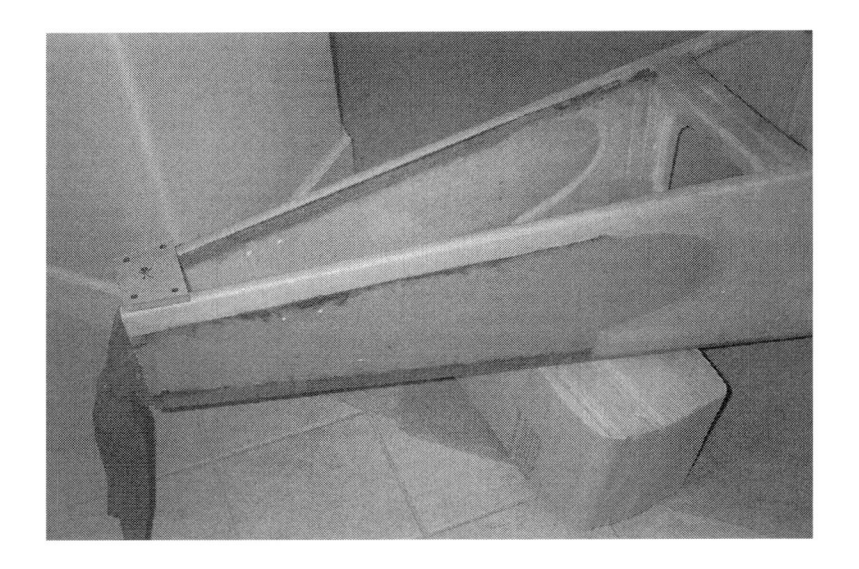

A tape measure is stretched from this tail pin to the wing spar edges on both left and right sides. This is done to lock in the yaw-axis alignment of the spar/wing to the fuselage centerline.

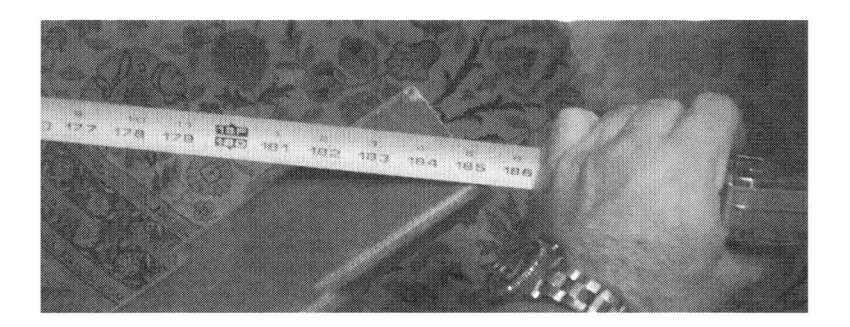

If you cannot tell already, this was all done inside my parent's house, as it was the only indoor space big enough to assemble a fuselage and wing spar setup indoors. This is one of the most critical alignments on the whole plane, so an indoor, take your time environment is a must have.

7.2 - Cleco Alignment

Clecos are not only for sheet metal fabrication, they work wonders with composites as well. Clecos are the ideal sheet holding and bond pressure tool. They can be installed with one hand from one side. When waxed, the three smooth sides of a Cleco are easy to release from a bonded joint. The spring back jaw pressure provides the clamping during a bonding cure operation. Just like sheet metal fabrication, use Silver 3/32 Clecos (#40 drill bit) for initial alignment and Copper 1/8 Clecos (#30 drill bit) for the final align and clamp. In initial setup and when installing Clecos for a bond, you need to work in a wheel-torque-like pattern; don't just go left to right. Typically, you pin a corner and check for warpage. If there is none, pin the next corner until all corners are pinned. Then go for the mid positions, then 1/4 positions, and so on.

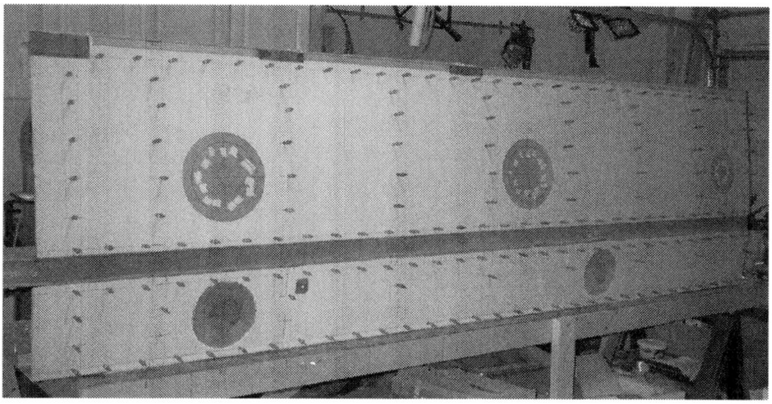

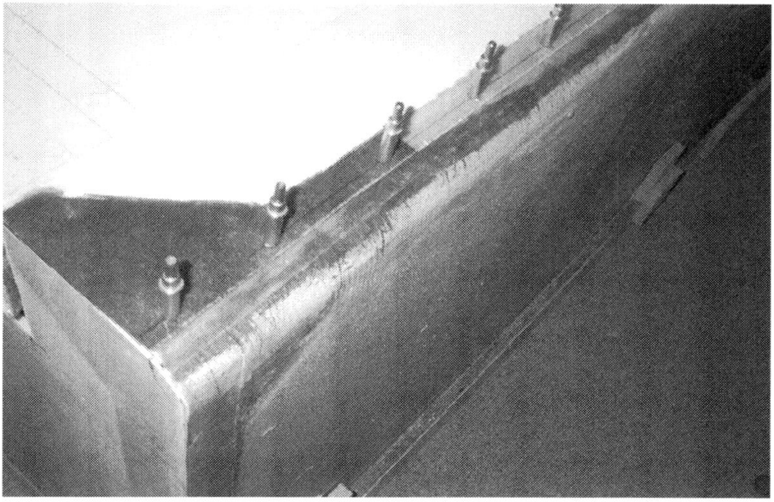

7.3 - Bonding Prep

The same basic steps are used for almost every bonded joint.

7.3.1 - Peel Ply Removal

Typically, you should use peel ply if you know an area is going to be bonded or will receive additional lamination on top of a cured layup. It's best to remove the peel ply once the laminate is cured rather than months or more down the road, as it can become more difficult to do later on. The peel ply should be peeled up. If it comes up in little pieces, use needle nose pliers. GIANT WARNING NOTE: **NEVER SAND AWAY PEEL PLY!**

One of my mentors told me of someone who decided to sand away the peel ply, rather than take the time to remove it correctly on a Lancair 4 project (over $100K in the airframe alone). After spending who knows how much on the project, including a good amount on a paint job, the wing tip was lifted up during inspection and the whole wing skin delaminated as the sanded peel ply acted like release agent. The airframe was scrapped and parted out. Remember: remove by peeling away, NOT sanding away.

7.3.2 - Mechanical Roughing Up.

Even with peel ply, you will still need to "de-flash" the surface. The three tools of the trade are:
#911 Dremel Aluminum Oxide Grinding Stone

This is a good go-to. The lack of sharp edges (like a cylinder) keeps it from digging in. I'm not aware of too many aftermarket options for this bit, but the Dremel version is under $5 and will last for months of use.
This contour also works well for prepping micro and Flox corner joints that were cured before a layup to ensure no high spots and a satisfactory surface for subsequent lamination.

1/2" Sanding drums, 60 - 100 grit all work well. These can be purchased in bulk and aftermarket through Amazon and e-bay. This tool is best used when you have free and clear access to an edge where the cylinder can run parallel to the surface and not dig in.

An alternative is to use a flexible shaft and the sanding drum. The output end of this device is around the same size as the drum, and will prevent the digging-in that would occur with the larger body Dremel and drum setup. They also sell air turbine powered rotary tools around the same size as a flex shaft.

The third sanding tool is 60-80 grit sandpaper on a foam sanding block. This tool is used to de-flash and achieve a dull appearance. Remember: Molds are shiny so the part can release because shiny surfaces are bad for a bonding and secondary layup operations.

Once all sanding is completed, make sure to use your brush attachment on the shop vacuum to pull up any loose dust. The brushes agitate the surface and help kick up anything that's not supposed to be there.

Final part of prep is to use isopropyl alcohol on a paper towel to wipe off anything that remains. Remember to apply the alcohol to the paper towel, and not the surface, so it can work correctly and allow the residual alcohol to quickly evaporate from the surface.

Make sure to wax all your Clecos used for bonding with mold release wax. Hi-Low Mold Release Wax P/N 01-00177 sold at Aircraft Spruce is a good go-to product for this. Extend the jaws in the Cleco, dip it in the wax, and wipe off the excess with a paper towel. Don't leave excess globs of wax behind on the Clecos as it can be left behind in large quantity on your bonded part.

7.4 - Bonding Trowel

Using pinking shears, you can cut the vinyl blind squeegees mentioned near the beginning of the book to give them a uniform peek and valley profile. This profile will evenly distribute the bonding adhesive just like a tile trowel and mortar.

Chapter 8 - Finishing

8.1 - General Game Plan

Some finishing steps can be incorporated in the initial layup and planning. In fact, some composite parts have the final paint as the first layer in the mold, and the plies are added to it after the fact. When popped out of the mold, cosmetically they are ready to go.

8.2 - Filling-in the Weave Options

For prepreg materials, there are surfacing films that are the first item down on the mold. Surfacing films offer a significant time and weight savings compared to filling in a pin-holed surface after the fact. For prepreg, wet layup, and RTM, you can use "veil cloths". These are light weight/high thread count/high resolution fiberglass cloths in the .5 - 3.5 oz/yd^2 range. Fiberglass Style 120 is a common option.

8.3 - UV protection

Composite structures can be resilient against nature. However, one of their weaknesses is ultra violet "UV" light, which causes the breakdown of the laminate matrix. Top and side surfaces are the most susceptible. The bottom side of a plane, boat, or car does not receive much UV exposure, so you usually don't need to worry about it. There are some targeted UV protection paints/coatings for composite structures. The simplest method is a black primer. Black absorbs visible light along with UV. To test the coating thickness, coat clear plastic with the protective coating and hold it up to the sun. If you don't see light penetrating through the primer, you're good to go.

Chapter 9 - Additional Resources

There are sources that provide a general overview of composite specifications. The official custodian has changed hands from the Department of Defense (DoD) to the private sector for many specs. This is common with many "Mil specs". As a practicing Aerospace Engineer for more than a decade, and working with the materials for over two decades, I have learned original Mil specs are the best bang for the buck. Almost all are free, and the only complaint is "copy of a copy" blurriness in the days of typewriters and copy machines.

Most of these documents originated in wartime (cold war included), and they contain the get-to-the-point, need-to-know information that will save the day. With the transition of specs to the private sector, good, solid technical content often gets dissolved or diluted into another spec (all pay ware of course) and filled in with Lawyer language. This is not to say the new specs are not without their merit, but realize that just because technical information was generated decades ago, does not mean it is without value.

Many Mil specs can be downloaded for free from: www.everyspec.com

Mil Handbook 17 (Mil HDBK 17) is one of the main composites go-to guides. There are five books, and 1-3 contain the main need-to-know information for most composites work. The primary purpose of MIL-HDBK-17 is the standardization of engineering data development methodologies related to characterization testing, data reduction, and data reporting of properties for polymer matrix composite materials.

Mil HDBK 691Adhesive Bonding: This handbook provides basic and fundamental information on adhesives and related bonding processes for the guidance of engineers and designers of military materiel. It should provide valuable information on most of the factors that must be considered in adhesive bonding, and should be of value in the preparation of specifications; including process specifications.

AC 43.13 FAA Advisory Circular (AC) Acceptable Methods, Techniques, and Practices – Aircraft Inspection and Repair. This resource contains methods, techniques, and practices acceptable to the Administrator for the inspection and repair of non-pressurized areas of civil aircraft, only when there are no manufacturer repair or maintenance instructions. This data generally pertains to minor repairs. The repairs identified in this AC may only be used as a basis for FAA approval for major repairs. You can download a free PDF copy from the FAA website www.faa.gov.

AC 20-107 Composite Aircraft Structure: This AC sets forth an acceptable means, but not the only means, of showing compliance with the provisions of Title 14 of the Code of Federal Regulations (14 CFR) parts 23, 25, 27, and 29 regarding airworthiness type certification requirements for composite aircraft structures involving fiber reinforced materials, e.g., carbon and glass fiber reinforced plastics. Guidance information is also presented on the closely related design, manufacturing, and maintenance aspects. The information contained herein is for guidance purposes and is not mandatory or regulatory in nature. This is also available as a free PDF from the FAA web site.

AGATE / NIAR / NCAMP

http://www.niar.wichita.edu/agate/

http://www.niar.wichita.edu/coe/ncamp.asp

http://www.niar.wichita.edu/

Composite Airframe Structures by Michael Niu has excellent reference information, including ABD matrix calculations. **ISBN-13:** 978-9627128069

Moldless Composite Sandwich Aircraft Construction by Burt Rutan. An excellent book for Long-ez style composite construction, including hot-wired blue polystyrene cores and other innovative methods. **ASIN:** B000BUJP5A

My personal Web site and Blog www.gouldaero.com is filled with various composite projects, products, concepts, and methods, along with other construction methods. www.gouldaero.com/pmc is the dedicated support page for this book.

About the Author

I have a diverse aeronautics experience, covering airframe and power plant design and development since the mid-90s. I am US Citizen and Embry-Riddle BSAE Alumni. My professional career has covered General Aviation, Commercial, Military Airframe and Engine OEMs.

I have been a member of the Experimental Aircraft Association (EAA) since 1993, and currently serve as a Technical Counselor in Composites and Firewall Forward, and Local Chapter VP. Beyond composites, I am also highly skilled at machining and TIG welding. In my professional Aerospace Engineering career, I have served as a Composites and Airframe Technical Fellow for aircraft gas turbine companies.

What truly kicked off much of the direction in my life was when my parents took me to Oshkosh 1993. If you have never been to Oshkosh or Sun-N-Fun, for just the cost of the admission ticket, you can get introductory hands-on/real-world training in every kind of major aircraft fabrication method, and I tried them all. Composite construction was the earliest method I was able to excel in, and I have had an exponential learning curve ever since then.

I was the Chief Engineer and Composite Fabricator on Jason Newburg's Race #68 Jamaica Mistaka, 1st place winner of the 2006 Reno National Air Races Silver Biplane Class.

I was the build partner on Jim Moore's Velocity SE four-seat all-composite canard aircraft.

6G / 95°F / 95% humidity Static Proof Load test on my very own, self-designed all-composite home-built aircraft.

This book is a practical summary of everything I have learned since 1992, along with the many "gotchas" that can hold you back from being successful with composites in a concentrated, need-to-know format.

I hope that you learned new things about composite constructions process, equipment, and setups with this book. Importantly, I hope to have given you the motivation to get into the skill, art, and science of composites.

Spencer Gould
EAA 466275
EAA Technical Counselor:
    Composites and Firewall Forward
Blog / Web Site: www.gouldaero.com
Composites Book support page www.gouldaero.com/pmc

For other DIY / home workshop books from this author, check out:

TIG Welding: GTAW need to know for beginners & the DIY home shop. 2nd Edition

ASIN: B06ZZCLYHY / ISBN 1973147009

TIG Book support page www.gouldaero.com/weld

Printed in Poland
by Amazon Fulfillment
Poland Sp. z o.o., Wrocław